The Library of Physics

CHARACTERISTICS AND BEHAVIORS OF WAVES

Understanding Sound and Electromagnetic Waves

April Isaacs

The Rosen Publishing Group, Inc., New York

For my sweet dad, the wave sensei

Published in 2005 by The Rosen Publishing Group, Inc.
29 East 21st Street, New York, NY 10010

First Edition

Library of Congress Cataloging-in-Publication Data
Isaacs, April.
Characteristics and behaviors of waves: understanding sound and electromagnetic waves / by April Isaacs.
   p. cm.
Includes bibliographical references and index.
ISBN: 978-1-4358-3722-5
1. Sound-waves. 2. Electromagnetic waves. I. Title.
QC243.I78 2005
539.2—dc22

                                                    2004013651

Manufactured in the United States of America

On the cover: Scientists can use computer voice recognition and speech synthesis software to study several voice patterns, which are the white jagged wave forms seen in this conceptual computer artwork.

Contents

Introduction   4

1 What Is a Wave?   7

2 Categorizing a Wave   14

3 Parts of a Wave   21

4 Behaviors of Waves   30

5 Waves in Our World   39

Glossary   43

For More Information   45

For Further Reading   46

Bibliography   46

Index   47

Introduction

Waves are everywhere, but you might not be conscious of them. Imagine that you're listening to your favorite band perform at a concert hall or chatting with a friend by telephone. Perhaps you've spent a summer vacation sunbathing by the pool, surfing along the coast of Hawaii, or flipping through radio stations during a long car trip. While you are doing each of these activities, you are encountering and utilizing waves. When you think of waves, perhaps you think of the waves in the ocean or doing "the wave" at a sporting event. These are all examples of waves and wavelike behavior that can help us to understand the nature of waves and the large role they play in the scientific perception of our world.

In the sixteenth and seventeenth centuries, scientists began to develop their theories based on experiments and mathematical analyses rather than on the observation of natural events. This period in history is sometimes called the scientific revolution. It was during this period that scientists became fascinated with the study of waves and began to see

Water waves that move at the surface of the ocean, such as this 70-foot (21.3 meter) wave on the north shore of Maui, Hawaii, are surface waves. Surface waves are unique because they are a combination of two wave movements—transverse and longitudinal—that create a circular wave movement in the water.

them as a means of unlocking the secrets of Earth and even our universe. Physicists have used their understanding of waves to develop sonar (sound navigation ranging), radio, and television and to improve how we communicate with one another through wireless Internet and cellular phones. Even some animals such as bats and whales use waves to navigate (to measure distances), to find food, and to locate other members of their species.

Because our daily lives are affected by and dependent on waves, it is important that we understand their characteristics and behaviors. For example, what happens when we listen to a radio program and how does that experience differ scientifically from when we listen to live music at a concert? The fact that waves exist everywhere in many different forms can make understanding them a bit overwhelming. Actually, waves are not as confusing as they may seem at first. Waves can generally be broken into two categories: those requiring a medium, such as air or water (referred to as mechanical waves), and electromagnetic waves that do not require a medium. This book will examine these two types of waves, their characteristics, the ways in which they behave, and their practical application in our daily lives.

What Is a Wave?

Waves are all around us, but what exactly happens when we encounter a wave? Some waves we can see, and therefore we can get an idea of their shape and how they behave. We can observe waves moving in an ocean or we can watch sports fans doing the wave in a stadium and actually see wavelike motion in action. However, there are some waves, such as seismic and sound waves, that we cannot see. We know they exist because we can hear them or sometimes feel their presence. There are waves that travel through liquids, solids, and gases, and some that can travel in a vacuum. There are waves that move horizontally and those that move vertically. How can we recognize something as being a wave? What are some of the basic attributes that are common to all types of waves? How do we know that a wave is a wave?

Making a Mechanical Wave

A mechanical wave can best be described as a disturbance that propagates, or travels, through a medium from one point to another. There is a simple

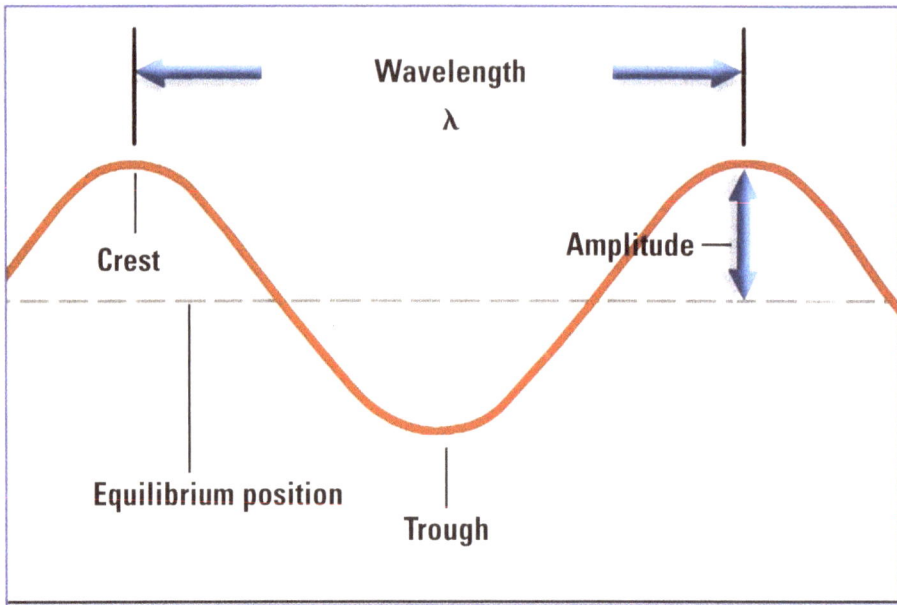

This illustration depicts the characteristics of a continuous wave. The highest point above the equilibrium position is called a wave crest. The lowest point below the equilibrium position is a wave trough. In a continuous wave, the distance between neighboring crests or troughs is always the same. The distance a wave travels in one cycle is called a wavelength.

experiment you can do at home or in the classroom to simulate a wave by using a piece of rope or heavy string that is about 12 to 15 feet (about 4 meters to 5 meters) long. Tie one end of the rope or string to the back of a chair or to a doorknob and hold the other end in your hand. Extend the rope or string as far as it will go and hold it out straight but with a little slack. When you hold it this way, there are no waves traveling through the rope. This is called an equilibrium position or a rest position.

First, let's illustrate a pulse. Move the rope or string quickly with your hand once, up and then down. You will notice that the rope or string also moves up, down, and then returns to its original rest position. This is a pulse. A pulse is a single disturbance propagating through a medium from one point to another.

Now move the rope or string end in a large, quick repeated up-and-down motion. By moving it this way with your hand, you're creating a periodic disturbance in the equilibrium position. You will notice a continuous motion of alternating hills and valleys that appear to move through the rope. You have now created a true wave! Unlike a pulse, a continuous wave is a periodic, or repeating, disturbance that travels through a medium from one point to another.

What Is a Medium?

Perhaps you've heard of artistic media, the materials that artists use to create paintings, sculptures, or photographs. For example, Paul Gauguin expressed himself through the artistic medium of painting, whereas Ansel Adams communicated his ideas through the medium of photography. A medium is a substance or a material through which something is transported from its source to another location. Imagine that you are in an art class and you want to create a likeness of an apple. You may choose to draw the apple with a pencil or to sculpt it from clay. In this case, the media are the pencil and the clay. Until your idea of how an apple should look travels through a medium (the pencil or the clay), it cannot

In art, a medium is the material or technique that an artist uses to produce a work of art. Top left: Commentators interviewed football player Reggie White for the television and radio media in 1997. Top right: This statue *Rocky* is an example of the medium of sculpture. Bottom: A third medium is shown, an oil painting by Paul Gauguin, entitled *The Flageolet Player on the Cliff* (1889). In physics, a medium is the substance, such as a rope or air, that is displaced as a wave travels through it.

be communicated. The same is true for many types of waves. A wave in the ocean travels through the medium of water. When your science teacher gives a lecture on physics, the sound of his or her voice travels through the medium of air.

Energy Transport Phenomenon

How does a wave maintain the energy to travel through the rope? How does it manage to move from the tips of your fingers to the doorknob? When understanding the propagation of waves, it is important to think of the medium as being interconnected particles, or parts that can interact with each other. Energy is sent from the source of a disturbance from one particle to its neighboring particle. Imagine a set of dominoes lined up next to one another on a table. When you tap the first domino, the force or energy that is sent from your finger to the first domino will transfer to the second domino and so on. When you tap a domino, you are transporting energy

Tsunamis

Tsunamis can be disastrous for those who live near the ocean. (The word "tsunami" is from the Japanese word for "harbor wave.") Tsunamis are a series of gigantic waves that only get bigger as they approach land and shallow water. They are caused by large-scale disturbances in the ocean such as earthquakes, impacts of meteorites, or volcanic eruptions.

A tsunami crashed on the California coast near Monterey. As a tsunami wave approaches the shore, the friction from the increasingly shallow ocean floor diminishes the velocity, or speed, of the waves. As the velocity decreases, the wavelengths become shorter and the heights considerably increase.

by displacing or moving the medium (the dominoes). However, when a wave transports a disturbance, it does so without displacing matter. After being tapped, the dominoes fall down. They do not return to their original upright position. When a wave passes through a medium, the medium always returns to its original rest position. This is called elasticity. Its effect is similar to the way in which a stretched-out rubber band can return to its original shape.

Elasticity is also sometimes referred to as an energy transport phenomenon. The fact that a wave can travel through a medium without permanently displacing matter is another defining characteristic of a wave. It is one way in which we can identify a wave.

Let's return to our rope experiment to better illustrate the energy transport phenomenon. When the rope is disturbed, the first part of the rope that is disturbed transfers its energy to the next part and then returns to its rest position and so on. When disturbed, the rope moves back and forth. It doesn't permanently freeze into the shape of a hill or valley. To get a better idea of how matter is not permanently displaced in a wave, consider the waves in the ocean. The wave that moves through the ocean does not displace its water. Instead, the medium (water) returns to its original position. If it did not return to a rest position, there wouldn't be any water left in the ocean! We know that we've found a wave when there is a continuous or periodic disturbance that travels through a medium from a source to its destination without displacing matter and leaves the medium essentially intact.

2 Categorizing a Wave

Now that we know how to recognize a wave, we need to learn how to distinguish one type of wave from another. Waves come in all shapes and sizes, and unless we have a logical way to categorize them, the study of waves can become quite confusing. Sometimes the difference between one wave and another can seem obvious. For example, an ocean wave is very different from a radio wave. The ocean wave travels through water whereas the radio wave does not require a medium to propagate itself. Yet these two types of waves have characteristics in common that make them waves.

Mechanical Waves and Electromagnetic Waves

Whether or not a wave requires a medium is the first step in categorizing it. In the introduction of this book, we learned that there are two general categories of waves: mechanical (sound is a type of mechanical wave) and electromagnetic waves. Electromagnetic waves can propagate through a vacuum (empty space). They do not require a medium to transfer a disturbance. Some examples of electromagnetic waves

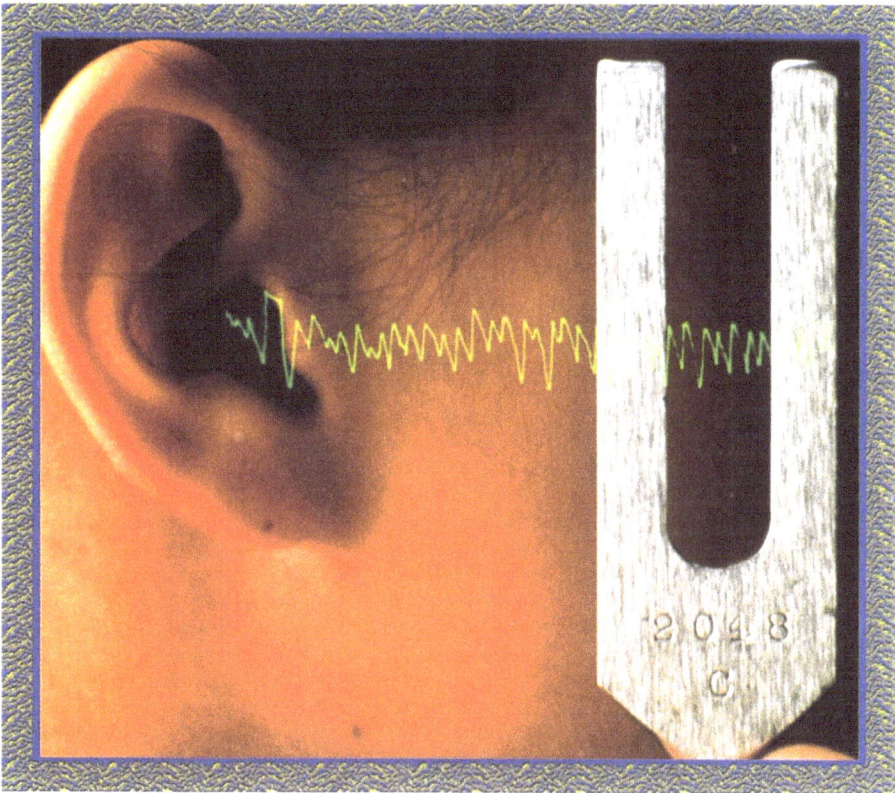

When struck, the prongs of a tuning fork vibrate, producing a sound wave. This wave travels from the fork to the outer part of the human ear, which directs some of the waves into the ear canal. Deep within the ear, the sound vibrations are perceived by nerves that send impulses to the brain. The brain finally interprets the vibrations as a musical note.

are radio waves, microwaves, X-rays, and light waves. We generally divide the electromagnetic range of waves into radio waves and light waves. The propagation of electromagnetic waves is also referred to as electromagnetic radiation. On the other hand, mechanical waves such as sound waves, require a medium to propagate. They cannot travel through a vacuum. If you were to stand on the moon and play a song on a violin, you would not hear the sound because you are standing in a vacuum. There is no air (or medium) on the moon for the sound to travel through. However, you are able to see the light from the sun because a light wave is able to transmit energy through empty space. Light waves and radio waves do not require a medium to propagate.

Now that you know how to determine if a wave is mechanical or electromagnetic, there are ways to categorize a wave even further. The next characteristic examines how a wave travels.

Transverse and Longitudinal Waves

Let's begin our study of wave movement by examining the direction of movement in mechanical waves.

Is Light a Wave or a Particle?

Since the seventeenth century, physicists have been arguing whether light is a particle or a wave. English physicist Sir Isaac Newton (1642–1727) was the first to declare light as a particle (made up of tiny molecules of matter). Christian Huygens (1629–1695), a Dutch physicist, argued that light traveled in waves and discounted Newton's theories. In 1900, German physicist Max Planck (1858–1947) introduced wave-particle duality (quantum theory), which stated that light waves behaved as both waves and particles.

In 1666, Sir Isaac Newton studied the colors that were produced when a ray of sunlight passed through a prism. He called the array of colors a spectrum.

Determining the direction of movement of mechanical waves is based on the elastic movement of the medium through which they are propagated. There are three main categories in which a mechanical wave can be classified: transverse waves, longitudinal waves, and surface waves.

If the particles of a medium move in a direction perpendicular to the direction in which the wave moves, the wave is transverse. Our experiment with the rope is an excellent example of a transverse wave. The wave you created was moving in a horizontal direction. It transferred energy from your hand to the doorknob. Meanwhile, the rope (medium) moved in a vertical direction. It temporarily displaced itself upward and downward in the shape of hills and valleys.

An audience wave is also a great example of how a transverse wave behaves. Suppose you are at a football stadium and one person begins an audience wave. That person begins by standing up and then sitting back down. As soon as she sits down, her neighbor stands up and so on. The chain continues from left to right, or vice versa, until the wave reaches the last person sitting on the bench. Each member of the audience (the medium) moves vertically (sits up and down), which is perpendicular to the horizontal direction of the wave (from left to right or from right to left). The important thing to remember about a transverse wave is that the disturbing or

Transverse wave

Amplitude

Wavelength

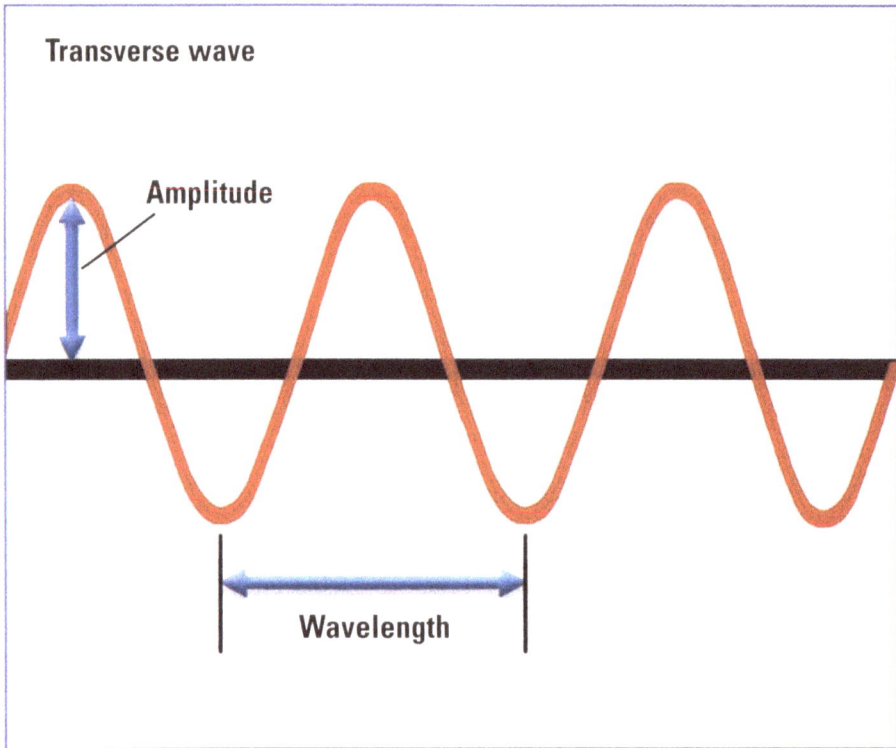

A transverse wave causes the particles of a medium to move perpendicularly to the direction in which the wave is moving. The amplitude of a wave is the maximum displacement from the rest or equilibrium position. The frequency of a wave is the number of complete vibrations per second measured at a fixed location, and it is measured in hertz.

actuating force is perpendicular to the direction of the wave. Transverse waves can be either mechanical or electromagnetic.

Another type of wave is a longitudinal wave. In a longitudinal wave, the particles of a medium move in a direction parallel to the direction in which the wave moves. Suppose a vibration is sent from a source to a particle in a medium. If it is a longitudinal wave, the particle will push its energy onto the neighboring particle. The particles of the medium are temporarily displaced from left to right in a horizontal direction while the wave is also moving left to right in a horizontal direction.

A sound wave is an example of a longitudinal wave. Imagine that you are standing across the room from a friend. Your friend says, "Hello, over there!" The sound makes its way across the room in a horizontal direction toward your ears. While the sound is traveling, it pushes air molecules against one another in a horizontal direction. Unlike the transverse wave, it doesn't move in the shape of hills and valleys. Instead, the sound wave moves in a repeating formation of compressions and rarefactions. Compressions are places in the medium that are pushed close together. Rarefactions are places in the medium that are spread apart.

Longitudinal waves can travel through solids, liquids, and gases. Liquids and gases do not support pure transverse waves. You might be wondering how an ocean wave is classified. Imagine that you are surfing on a hot sunny day. The wave you're riding might seem to be a transverse wave. You're surfing on a wave that looks as if it's moving in a vertical direction perpendicular to the direction of the wave. In other words, the wave is sending you shoreward as the water is moving upward and downward. However, because transverse waves cannot travel through liquids and gases, an ocean wave must be categorized as something else. Although there are longitudinal waves in the depths of the ocean, the

Longitudinal wave

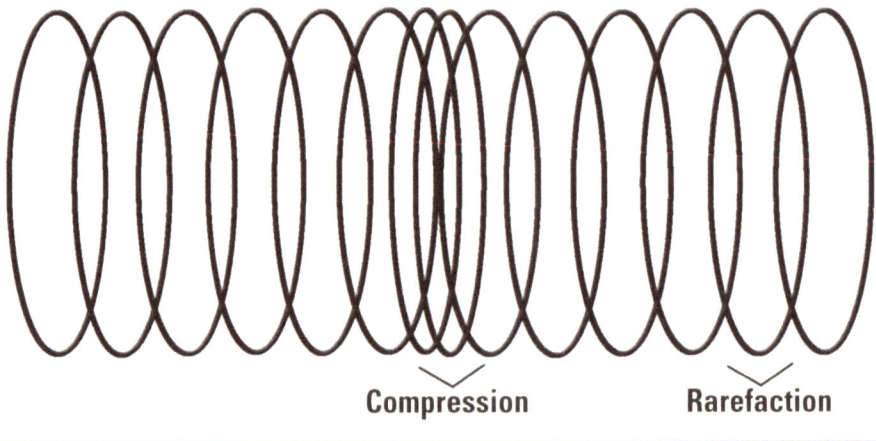

Compression   Rarefaction

A longitudinal wave causes the particles of a medium to travel parallel to the direction of the wave. Sound waves are longitudinal waves. Compression is the area of a longitudinal wave in which the density and pressure are greater than normal, whereas rarefaction is the region of a longitudinal wave in which the density and pressure are less than normal.

waves that break on the ocean's surface are known as surface waves. A surface wave is a combination of transverse and longitudinal movement that creates circular motion of the medium (water). Only the particles at or near the surface of the medium have this circular motion. The motion of the particles tends to decrease as you move farther from the surface.

3

Parts of a Wave

In the previous chapter, we covered transverse waves and compared their upward and downward motion to alternating hills and valleys. These alternating hills and valleys are known as crests (hills) and troughs (valleys). A crest is the point of maximum upward (or positive) displacement along a wave. In other words, it's the maximum point at which the medium is displaced upward from the rest position. The trough is the point of maximum downward (or negative) displacement along a wave. That is to say, a trough is the maximum point at which the medium is displaced downward from the rest position.

The crests and troughs represent the presence of a disturbance in a medium. Earlier, we learned that when a rope is held out straight in a rest position, there are no waves present in the rope. Once your hand moved the rope up and down, you noticed that a disturbance was sent through the length of rope. But how do we measure this disturbance that is moving through the rope? If you move the rope by making short up-and-down motions with your hand, the crests and troughs in the rope are smaller. However, if you move the rope by making long up-and-down motions with your hand, the crests and troughs in the rope are larger. To measure the level of a disturbance in a transverse wave, we measure the height of the crest (or the depth of the trough) from the rest position. This distance is known as the amplitude of a wave. The amplitude is one way of measuring a wave.

Amplitude and Wavelength

To determine a wave's amplitude, the maximum distance the wave moves from its equilibrium or rest position must be measured. The wavelength (or length of one complete wave cycle) of a transverse wave is the distance from one crest to the next crest or from one trough to the next trough. The wavelength is sometimes represented by the Greek symbol λ (lambda).

So how do we measure the amplitude of a longitudinal wave, which has compressions and rarefactions rather than crests and troughs? In a longitudinal wave (for example, a sound wave), the amplitude is the maximum distance a particle is pushed (due to compression) or pulled (due to rarefaction) from its equilibrium position. This is usually measured as the amount of deviation of pressure above and below atmospheric pressure. A unit of pressure is called a pascal. There are different types of instruments that are used to measure pressure. An example of an instrument that measures air pressure is a barometer. It can measure slight changes in atmospheric pressure that indicate changes in weather.

When dealing with sound relative to the human ear and its ability to measure loudness, scientists found that the human ear can determine a wide range of sound levels. Scottish-born American scientist Alexander Graham Bell (1847–1922) worked with people who were deaf or who had hearing impairments. He developed a system for measuring loudness that enabled him to easily determine the range of a person's hearing. The unit of measurement in this system was eventually named for him and called the bel. A decibel is equal to one-tenth of a bel and is denoted as dB. One example of how this sound-measuring system is used is the following: a reference

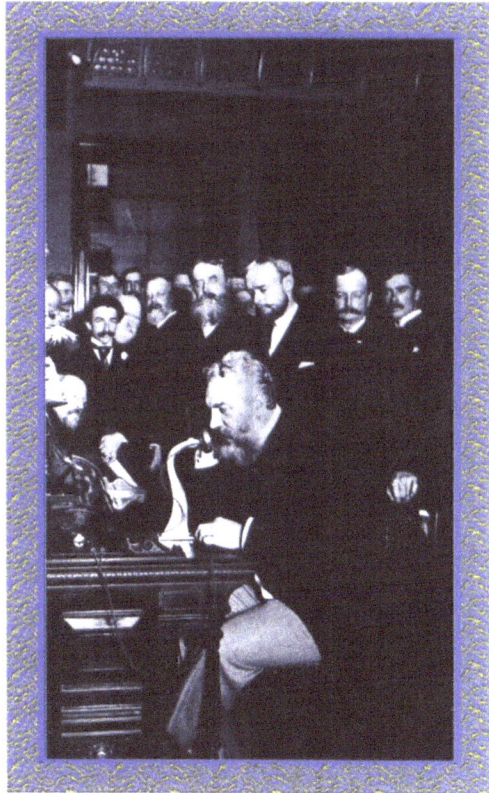

In 1892, Alexander Graham Bell opened the 932-mile-long (1,500-kilometer-long) telephone line between New York and Chicago. It was the world's first automatic telephone exchange and had more than 100 extensions.

sound level corresponding to a pressure of 20 micropascals (20 x 10^{-6} Pa) is designated as 0 dB. (This intensity in energy flow per unit area is the value 1x10^{-12} watts/m^2.) This is a sound level that a person with normal hearing can just barely hear. A sound level that is ten times louder than this reference sound level is 10 dB. A sound level that is 100 times as loud as the reference sound level is 20 dB, and something 1,000 times as loud as the reference sound level is 30 dB. (The decibel scale measures sound intensity. It is logarithmic, so each increase of 10 dB means that the sound is ten times more intense or louder.) For reference, a sound level in the range of 120 dB to 140 dB is considered to be the threshold of pain. Following our example, then, this threshold of pain corresponds to a sound level that is about 100,000,000,000,000 (100 trillion) times as loud as the reference level. Consequently, the dB system is a handy way to avoid dealing with large numbers. Bell's work with sound eventually led him to the development of his most famous invention, the telephone.

When you measure a wave's amplitude, you are measuring the level of disturbance in a medium. If you gently tug on the rope in your experiment, the disturbance is low. If you forcefully yank on the rope, the disturbance will be high. The magnitude (the height or depth) of a disturbance is directly related to the level of force that is used to introduce that disturbance. For example, if you whisper a secret to your classmate during class, the initial force of disturbance in the sound wave is low. However, if you are driving down the street, with music blaring through your state-of-the-art speakers, the disturbance that is traveling through the medium (the air) is high.

Whenever you measure the amplitude of a mechanical wave, you are usually measuring its displacement in units of length (that is, in inches, feet, meters, etc.) or in pressure (in pascals or decibels). The displacement can be transversal (perpendicular to the direction of wave travel) or longitudinal (in the direction of wave travel). Electromagnetic waves do not require a medium to propagate, whereas mechanical waves do. Therefore, when measuring the amplitude of an electromagnetic wave, we measure its electric field or the strength of its radiation. For light waves, which are electromagnetic waves, the amplitude can be linked to light intensity. For example, the intensity of light from a common lightbulb is measured in units of lumens. The amplitude of radio waves can be related to radiated power. The strength of a radio station near its transmission tower is measured in kilowatts (units of 1,000 watts). However, when the radio waves reach a radio, they are greatly weakened and are on the order of milliwatts (units of 1/1,000 of a watt).

Period and Frequency

To measure wavelength, we measure the distance of one repetition in a wave cycle. Similar to wavelength, a period is a measurement of the time that it takes to complete one wave cycle. To measure how often a

High frequency wave

Pressure

Time

Period

Low frequency wave

Pressure

Time

Period

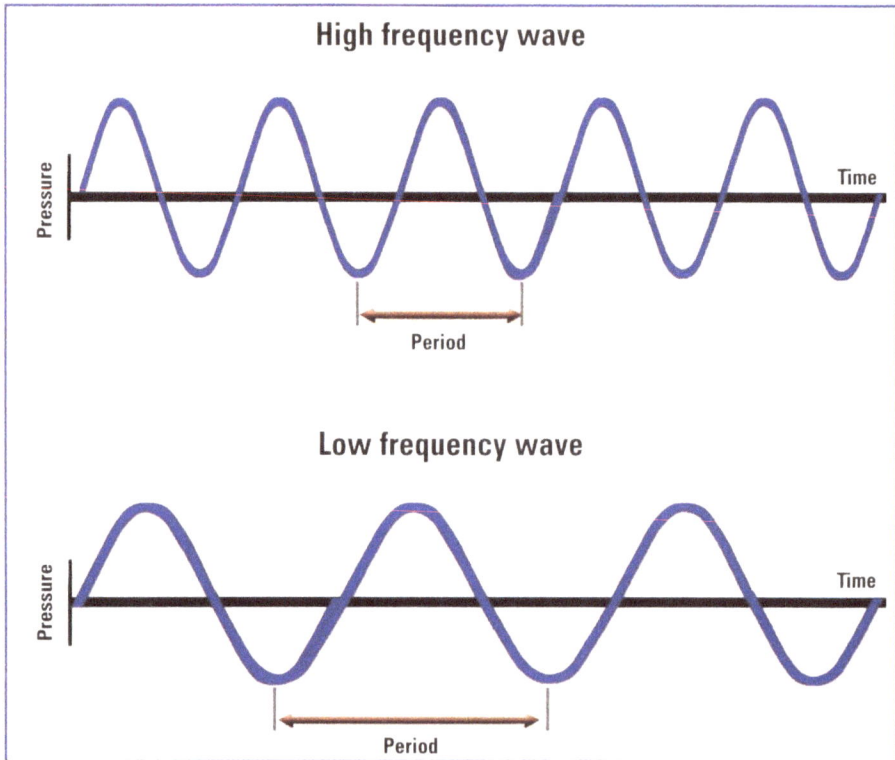

The frequency of sound waves is measured by the number of back-and-forth vibrations of the particles per unit of time. A sound wave with a high frequency has a pressure time plot that would have a small period, or a small amount of time between successive points of high pressure. A sound wave with a low frequency, then, would correspond to a pressure time plot with a large period, or a large amount of time between consecutive points of high pressure.

cycle occurs in a wave, we measure the wave's frequency. If a cycle is repeated many times in a time interval (for example, a second), the wave has a high frequency. If a cycle does not have many repetitions, it has a low frequency. Imagine the rope and the doorknob experiment again. Suppose it took exactly one second for the wave to travel from your hand to the doorknob. The number of crests (or troughs) that occur in one second represents the wave's frequency, or cycles per second. For example, AM radio waves have a low frequency whereas microwaves operate at a higher frequency. Sound waves can also have a wide range of frequencies. Bats and dogs can detect high-frequency sound waves whereas the human ear can detect only lower-frequency sound waves.

Frequency and the Electromagnetic Spectrum

The unit of frequency (one cycle per second) is commonly referred to as one hertz (1 Hz). The hertz is named for the German physicist Heinrich Rudolf Hertz (1857–1894), who, in the 1880s, was the first to transmit and receive radio waves. Radio waves are a form of electromagnetic waves. They do not require a medium to propagate. Hertz based his experiments on the theories of Scottish physicist James Clerk Maxwell (1831–1879), who postulated mathematical relationships between electricity and magnetism, thus founding the science of electromagnetics. Hertz used these findings to prove that the speed of radio waves in a vacuum is equal to the speed of light. The speed of light in a vacuum is 299,792,458 meters per second, or approximately 186,282 miles per second. If an object could travel at the speed of light it could circle Earth at the equator seven times in one second. He also demonstrated how magnetic and electric fields could detach themselves from a medium and travel through a vacuum. He called these new waves hertzian waves.

During experiments from 1886 to 1889, Heinrich Rudolf Hertz discovered that electromagnetic waves are long, transverse waves that travel at the speed of light and can be reflected, refracted, and polarized like light. We now call these waves hertzian or radio waves, and the unit of frequency, the hertz, is named after him.

Although Hertz was the first to transmit and receive radio waves, messages weren't sent across radio waves until 1901 when a young Italian scientist, Guglielmo Marconi (1874–1937), read about Hertz's waves and became astounded by them. He wanted to find a way to send messages through them. In 1901, Marconi publicly announced that he had sent a radio signal across the Atlantic Ocean from Cornwall, England, that was received in Newfoundland, Canada. Since the discovery of radio waves, our lives have drastically changed and improved.

Radio waves are a form of electromagnetic radiation (an electromagnetic wave). They behave in ways that are similar to those for light, X-rays, gamma rays, and microwaves. However, to differentiate the types of electromagnetic waves, scientists have categorized them according to their frequencies, wavelengths, and energy. The waves are then classified into what is known as the electromagnetic spectrum.

How do we measure the frequency of an electromagnetic wave? In the case of a radio wave, we can measure the time it takes to travel between peaks in the electric field (similar to the distance between crests in a sound wave). For example, in one second, we can count the number of peaks that occur, or cycles per second, as the number of hertz. Radio engineers can measure the radio frequency by using a tunable filter that is sensitive to energy at a given frequency determined by the selected tuning position. The device that measures radio frequencies is called a spectrum analyzer. In an earlier discussion, we learned that the time it takes a wave to travel between crests or troughs is called the period of the wave. The distance the wave travels in one period of time is called the wavelength. Because radio waves can travel very quickly, they can cover a lot of distance in a short period of time. At relatively low radio frequencies, for instance, 3 kHz (3 kilohertz, or 3,000 periods per second), the wavelength is measured in meters. At extremely high radio frequencies, for example, around 1,000 million Hz (sometimes called a gigahertz or GHz), the wavelength is measured in centimeters. Since light waves occur at even higher frequencies, their wavelengths can be very short and are therefore measured in nanometers (one-billionth of a meter).

4 Behaviors of Waves

Now that you know what a wave is, the different parts of a wave, and how to categorize a wave, you might be wondering exactly how a wave behaves. For instance, what happens during your rope experiment when the wave reaches the doorknob? In other words, what happens when a wave runs out of a medium?

Where one medium ends and another medium begins is known as the boundary. How a wave reacts when it encounters a boundary is described as the wave's boundary behavior. Imagine yourself as a wave. You are walking down the sidewalk when you bump into a freestanding door. What do you do? Essentially, you have one of three choices. You can turn around and go back the way you came, you can open the door and walk through it, or you can walk around it to get to the other side. When a wave encounters a boundary, it has the same basic choices. It can either bounce off the boundary, pass through it, or skirt around it. These different reactions (or behaviors) are known as reflection, refraction, and diffraction.

Making a Ripple Tank

There is a common experiment you can do in your science classroom or at home to exhibit the behaviors of waves. It is called a ripple tank experiment. A ripple tank can be constructed from readily available household items. You will need the following supplies:

a) A Pyrex (glass) baking dish, measuring approximately ten inches by sixteen inches (twenty-five by forty-one centimeters). This kind of dish is usually used for making sheet cakes.

b) A sheet of white paper measuring somewhat larger in size than the Pyrex dish. If you can't find a

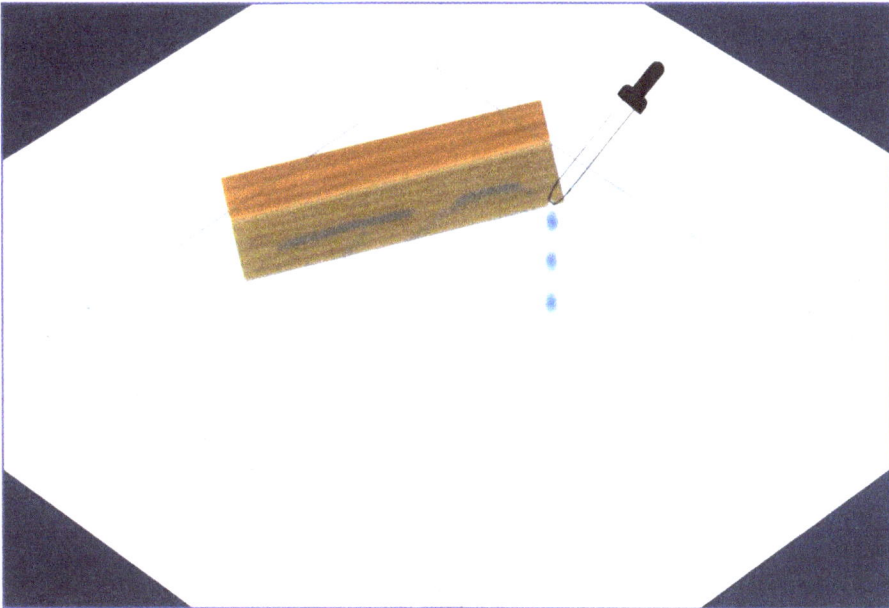

A ripple tank can be used to study the behaviors of waves as they move from one medium to another. By using a ripple tank, we can demonstrate wave behaviors such as reflection, refraction, and diffraction. The drops from an eyedropper cause the disturbance, the water in the dish is the medium, and the wooden board acts as the barrier.

single sheet this big, you can tape several pieces of standard 8.5-inch by 11-inch typing paper together.

c) A block of wood, measuring approximately 1 by 2 by 9 inches (2.5 x 5 x 23 cm) that can be used as a barrier.

d) An eyedropper or medicine dropper for creating waves.

Place the sheet(s) of paper on a table or a kitchen countertop and then place the Pyrex dish on the paper. Make sure there is proper lighting to view the dish. Next fill the glass dish with water until it is about half full. Then place the wood block in the dish to act as a barrier. You may have to hold the wood in place or put a weight on it to keep it from floating away. Now you are ready to begin using the ripple tank!

Reflection

When you imagined yourself encountering a free-standing door on the sidewalk, one of your choices was to turn around and go back the way you came. In the study of the behavior of waves, this is known as reflection. Let's return to our rope experiment once again. Create a pulse in the rope by yanking it once. The rope will move up and down until it hits the doorknob and then returns to your fingers. Some of the wave has been reflected back through the rope, but some of it has also passed through the doorknob (this is known as refraction, which will be discussed later in the chapter).

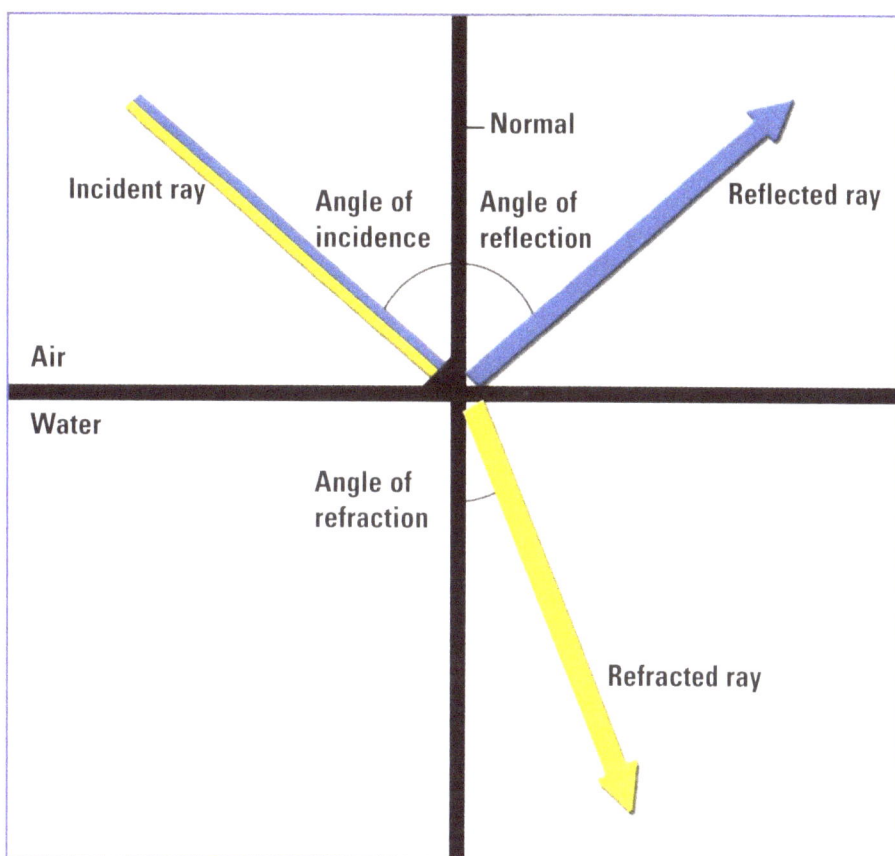

Wave properties of reflection and refraction are illustrated in this ray diagram. The law of reflection states that the angle of incidence is equal to the angle of reflection. In this diagram, the angle of incidence is the angle between the incident ray and the barrier, labeled here as the normal.

Now let's demonstrate reflection in our homemade ripple tank. The piece of wood will act as a barrier. Lay the wood across the width of the Pyrex dish. Use drops from your eyedropper to create a disturbance. A few waves will be sent out in the water. Once they hit the piece of wood, you will see them bounce off the wood and head in the direction of their initial disturbance. Now you have witnessed reflection in action. Reflection occurs when a wave reflects, or bounces, off the barrier and moves in a different direction. (Reflection can also occur when waves pass between media of different properties, such as density.) One important rule to remember when studying wave reflection is that the angle in which a wave approaches a barrier is always equal to the angle in which it is reflected. This principle is called the law of reflection. It is commonly applied to light waves that reflect off a surface.

Refraction

You learned that when a wave reflects, it bounces off the boundary and travels in a direction that is different from its original one. When a wave is refracted, it also changes direction. However, unlike reflected waves, which bounce off a boundary, refracted waves change direction after passing through a boundary (that is, when they pass from one medium to another medium). When a wave is refracted into a different medium condition, its speed and wavelength change. A tsunami is an example of this behavior. It begins as the result of a large-scale disturbance in deep waters. As the tsunami hits shallower water, it changes direction and moves shoreward. As the tsunami wave moves shoreward, its speed and wavelength decrease while its amplitude increases.

To demonstrate the way that waves change speed when refracting, you can use your ripple tank. Place in it a thin sheet of wood or plastic about two to three inches (5 cm to 8 cm) thick. It should be wide and long enough to take up one-third of the tank. The idea is to make one end of the ripple tank shallower than the other. At the tank's deep end create waves by dropping water with the eyedropper. As the waves reach the tank's shallow end, they appear to slow down.

Diffraction

When a wave undergoes diffraction, it doesn't pass through one medium to another or bounce off a boundary. Instead, a diffracted wave skirts around a barrier that is in its way. When a wave is diffracted, its level of diffraction (or the force it takes to bend around a corner) either increases or decreases along with its wavelength. An example of diffraction in sound waves is if your mother is in the kitchen and yells toward you in the next room that dinner is on the table, her voice is being diffracted. The wave diffracts from, or deflects from, the walls that separate the kitchen from the next room and travels to your ear.

To demonstrate diffraction in your ripple tank, place two pieces of wood in the water, leaving a small gap between them. When you send a disturbance through the water with your dropper, the waves travel around the barriers.

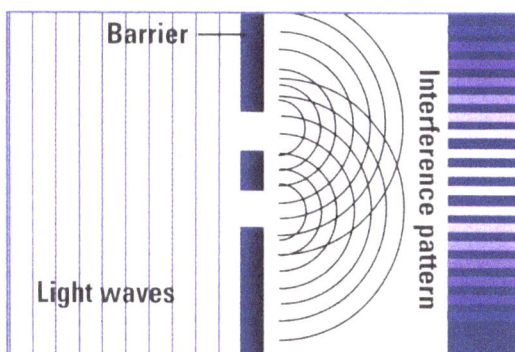

Light waves are diffracted at two openings in this barrier. Circular waves are formed at each opening. When these waves interfere with one another, the interference pattern can be projected onto a screen and results in alternative light and dark bands.

Interference

Suppose that two separate waves meet inside the same medium. What would happen? For example, when a wave meets another wave traveling in an opposite direction in your rope, the occurrence is called interference. There are two types of interference: constructive and destructive. In constructive interference, the crests and troughs of two waves are in phase, or coincide, causing their amplitudes to

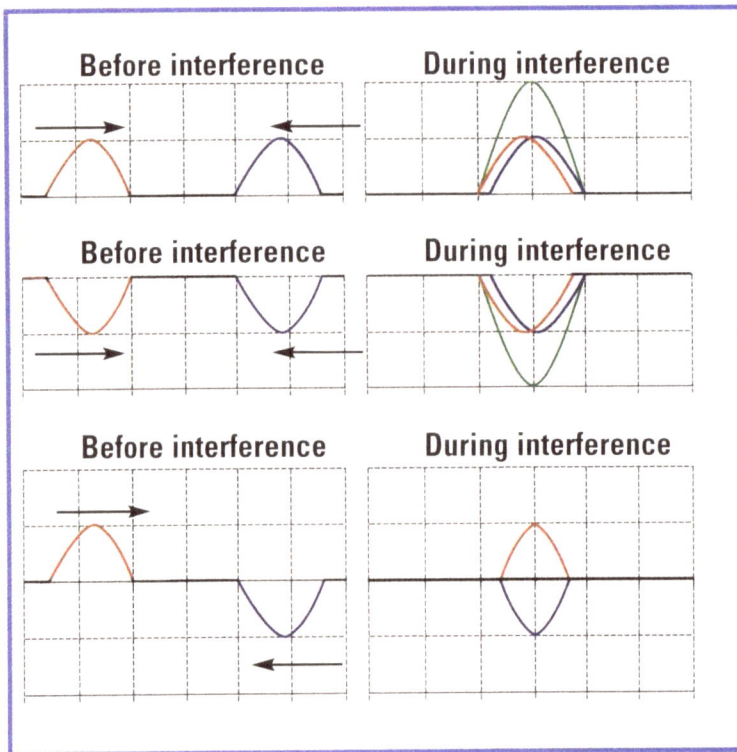

Before interference	During interference
Before interference	During interference
Before interference	During interference

At the top left are two waves traveling before interference and at the top right are the crests of the two waves as they overlap. The resulting shape of the displacement of the medium when the waves meet is drawn in green. This shows constructive interference. The second set of diagrams shows the constructive interference of two troughs meeting. The third set shows destructive interference, where two interfering waves have the same maximum displacement in the opposite direction. At the instant of complete overlap, there is no disturbance.

increase. In a destructive interference, the crest of one wave lines up with the trough of another. When a crest is lined up with a trough, meaning that they have displacements in the opposite direction, the crest and the trough cancel each other out and the medium returns to the rest position.

The Doppler Effect

Have you ever listened to a weather program on television and heard the term "Doppler radar"? Weather stations use Doppler radars to determine the velocity (or speed) of an approaching storm. By determining how fast a storm is approaching, the weathercaster can predict—with a certain level of accuracy—when each neighborhood can expect the storm to hit. But how does Doppler radar determine the speed of a storm? Doppler radar works on the principles of the Doppler effect. The Doppler effect is an evident change in the frequency or wavelength of a wave that is perceived, or noticed, by an observer moving relative to the source of the wave. With a Doppler radar, the storm's speed is measured in relation to a specific geographic location.

An example that is commonly used to demonstrate the Doppler effect in action is that of the sound of sirens coming from a moving police car or ambulance. If you are sitting on your porch as a police car drives past, you will notice as it approaches you that the sound gets louder, but also its pitch or frequency

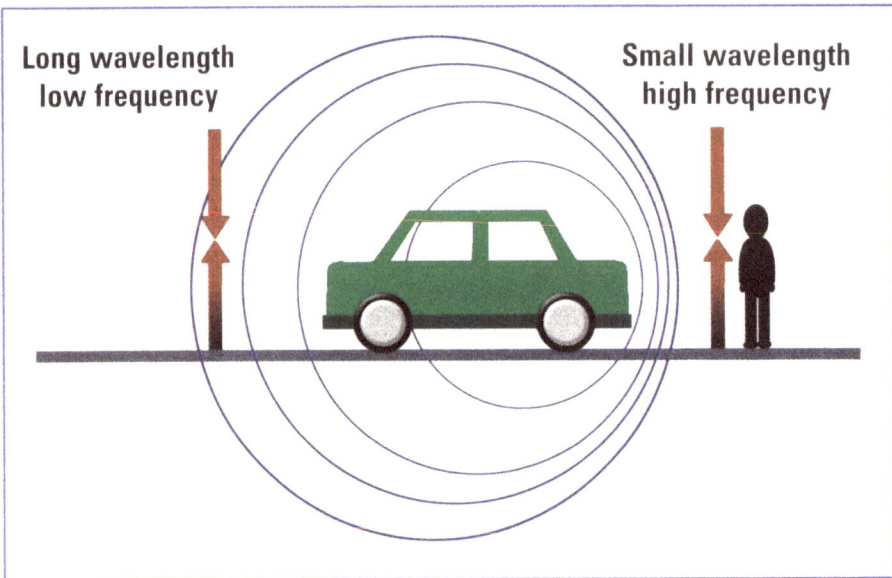

In the Doppler effect, if a car's siren is moving toward you, the frequency seems to increase as the sound approaches you. When the siren moves toward you, more sound waves are crowded into the space between you and the siren. The siren's frequency doesn't change, but you perceive a different frequency because of the relative motion between you and the siren.

increases. As the police car passes by you its sound level decreases and its pitch or frequency gets lower. However, the sound's frequency is not changing. As the police car moves away from you, its driver (who is moving with the disturbance) still hears the siren at the same frequency. As the police car approaches you, it decreases the wavelength between you and the initial disturbance. As the police car passes you, it increases the wavelength between you and the initial disturbance.

Although the Doppler effect is easiest to explain and notice in sound waves, it is a behavior that can exist in any wave, either a mechanical or an electromagnetic wave.

5 Waves in Our World

Throughout this book, we've defined a wave, its different parts, and how waves behave. Now that you have a better understanding of the characteristics and behaviors of waves, let's take a look at some ways that scientists use the study of waves to further our scientific understanding of Earth and the universe.

Sonar

If you have ever seen a movie or read a book about submarines, you are probably familiar with the term "sonar" (sound navigation ranging). Sonar is a method used by submarines and ships to navigate through the ocean or detect other submarines. When using sonar, a submarine sends out a ping (a disturbance) and waits to see if the ping bounces back (reflects) from a barrier. The distance between the submarine and the barrier is determined by measuring the time it takes the pulse (caused by the ping) to return. Using a ping to cause a wave to reflect off of a barrier (perhaps another submarine) is called active sonar. There is also another type of sonar called passive sonar. Passive sonar listens for waves without causing an initial disturbance (pinging). For example, passive sonar picks up sound waves that are sent out by whales or other submarines using active sonar. Sonar is used on many ships to help them avoid collisions with objects below the surface of the sea.

Seismology

Seismology is an earth science and a wave-related field. Seismologists study earthquakes, or the

A seismologist studies a seismograph that measured the wave activity from an earthquake that happened in western Mexico on September 30, 1999. Seismometers record the types of waves that cause earthquakes. Primary waves push and pull the earth, whereas secondary waves move the earth from side to side.

movement of waves through Earth's surface. By determining how seismic waves behave, seismologists have been able to discover that Earth's outer core is made of liquid. This was one of the first and most important discoveries made in seismology. Seismologists discovered that only longitudinal waves could travel through Earth's outer core. As you learned in an earlier chapter, transverse waves cannot travel through liquid or gas. However, the waves produced by an earthquake are both transverse and longitudinal. Yet, only the longitudinal waves could travel through Earth's outer core. This discovery led scientists to conclude that Earth's outer core is made of a liquid.

Radio Astronomy

One of the newer fields in astronomy is known as radio astronomy, or the transmitting and receiving of cosmic signals. In radio astronomy, scientists measure and observe the characteristics of radio waves they receive from outer space. Since the 1930s, radio astronomy has greatly increased our knowledge of the universe. Through radio astronomy, scientists have discovered previously unknown stars and constellations, and have been able to learn more about the presence of dark matter in the universe.

Because radio astronomers require gigantic antennae to receive radio signals from outer space, the United States built an observatory and laboratory for radio astronomers, called the National Radio Astronomy Observatory (NRAO). The NRAO operates

The Robert C. Byrd Green Bank Telescope is covered with a reflecting surface of aluminum foil that enables scientists to observe wavelengths of 1 inch (2.5 centimeters) and more. The telescope began operating in 1965, and it is sometimes used by the Search for Extraterrestrial Intelligence Institute.

the world's largest, entirely steerable radio telescope, the Robert C. Byrd Green Bank Telescope (GBT). The GBT is a 328-foot (100-meter) telescope and is located in Green Bank, West Virginia.

Sonar, seismology, and radio astronomy are just a few of the ways in which scientists have used the study of waves to further our understanding of the universe and improve our daily lives. Imagine how different your life would be had we never discovered waves! You wouldn't be able to call your friend on your cell phone, or from any phone for that matter. You wouldn't be able to watch television. Waves are all around us and they affect our daily lives. Although we have improved our lifestyles and increased our knowledge of our world through the study of waves, there is still much more to learn.

Glossary

amplitude (AMP-lih-tood) The distance from a crest or trough (or compression or rarefaction) to the rest position. In the case of a stretched rope or string, amplitude is measured in meters, whereas for sound waves, amplitude is measured in units of pressure.

compression (kum-PREH-shun) A place in a wave in which the particles of a medium are pushed together. A compression is the opposite of a rarefaction.

crest (KREST) The point of maximum upward (hill-shaped) displacement along a wave. A crest is the opposite of a trough.

density (DEN-sih-tee) The closeness of mass in a unit of a substance.

deviate (DEE-vee-ate) To turn away from an original course or form.

diffraction (dih-FRAK-shun) A behavior in which a wave travels around a barrier.

disturbance (dih-STUR-bents) An initial displacement of energy in a medium.

elasticity (ih-lass-TIH-sih-tee) A phenomenon in which energy is transported without permanently displacing matter. Also known as the energy transport phenomenon.

electromagnetic spectrum (ih-lek-tro-mag-NEH-tik SPEK-trehm) The spectrum containing all the different kinds of electromagnetic waves, ranging in wavelength and frequency.

electromagnetic wave (ih-lek-tro-mag-NEH-tik WAYV) A type of wave that does not require a medium to propagate. Light and radio waves are electromagnetic waves.

equilibrium (ee-kwih-LIH-bree-em) The rest position of a medium.

frequency (FREE-kwen-see) The number of times a cycle occurs during a wave.

law of reflection (LAHW OV rih-FLEK-shen) A scientific law that states that the angle in which a wave approaches a barrier is always equal to the angle in which it is reflected.

longitudinal (lahng-jih-TOOD-ih-nel) A wave characteristic in which a wave travels in a direction that is parallel to the direction of the movement of the medium.

magnitude (MAG-nih-tood) A greatness in size or amount.

mechanical wave (mih-KA-nih-kul WAYV) A type of wave that requires a medium to travel from one point to another. A mechanical wave cannot move through a vacuum. Sound waves are a kind of mechanical wave.

medium (MEE-dee-um) A substance or a material through which a machanical wave can travel.

pascal (pas-KAL) A measurement for one unit of pressure.

propagate (PRAH-pah-gayt) To travel or move from one place to another; to relocate.

radio astronomy (RAY-dee-oh uh-STRAH-nuh-mee) The study of radio waves in space.

rarefaction (rar-eh-FAK-shun) A place in a wave in which the particles of a medium are spread apart. A rarefaction is the opposite of a compression.

reflection (rih-FLEK-shen) A behavior in which a wave bounces off a barrier.

refraction (rih-FRAK-shen) A behavior in which a wave passes through a barrier.

seismology (syz-MAH-leh-jee) The study of earthquakes and seismic waves or waves that move through Earth's surface.

sonar (SOH-nar) An abbreviation for sound navigation and radar, a method commonly used by navies and other military forces in which sound waves are used to navigate underwater and locate submarines.

transverse (trans-VERS) A characteristic in which the particles of a medium are displaced perpendicular to the direction of the movement of a wave.

trough (TROFF) The maximum point of downward (valley-shaped) disturbance along a wave. A trough is the opposite of crest.

vacuum (vah-KYOOM) Empty space.

wave (WAYV) A disturbance that travels from one point to another without permanently moving matter.

wavelength (WAVE-leynkth) One complete cycle of a wave. Wavelength is determined by the measurement from crest to crest, trough to trough, or from compression to compression or rarefaction to rarefaction.

For More Information

American Physical Society
One Physics Ellipse
College Park, MD 20740-3844
(301) 209-3200
e-mail: letters@aps.org
Web site: http://www.aps.org

Web Sites

Due to the changing nature of Internet links, the Rosen Publishing Group, Inc., has developed an online list of Web sites related to the subject of this book. This site is updated regularly. Please use this link to access the list:

http://www.rosenlinks.com/liph/chbw

For Further Reading

Baierlein, Ralph. *Newton to Einstein: The Trail of Lights: An Excursion to the Wave-Particle Duality and the Special Theory of Relativity.* New York: Cambridge University Press, 2001.

Feynman, Richard Phillips. *Feynman Lectures on Physics.* Reading, MA: Addison-Wesley Publishing Co., 1970.

Flatow, Ira. *They All Laughed . . . From Light Bulbs to Lasers: The Fascinating Stories Behind the Great Inventions That Have Changed Our Lives.* New York: Perennial, 1993.

Gardner, Ralph. *Light, Sound and Waves Science Fair Projects: Using Sunglasses, Guitars, CDs and Other Stuff.* Springfield, NJ: Enslow Publishers, 2004.

Bibliography

Elmore, William E., and Mark A. Heald. *Physics of Waves*, Mineola, NY: Dover, 1985.

Feynman, Richard Phillips. *QED*. Princeton, NJ: Princeton University Press, 1988.

Kraus, J. "Antennas Since Hertz and Marconi," *IEEE Antennas & Propagation Magazine* 33 (1985): pp. 131–137.

Silverman, Mark P. *Waves and Grains.* Princeton, NJ: Princeton University Press, 1998.

Sinclair, Jim. *How Radio Signals Work.* New York: McGraw Hill-TAB Electronics, 1998.

Index

A
amplitude, 22, 24, 25, 36–37

B
Bell, Alexander Graham,
    23–24
boundary, 30, 34
boundary behavior, 30

C
continuous wave, 9

D
decibel, 23–24
diffraction, 30, 35
Doppler effect, 37–38

E
electromagnetic radiation, 28
electromagnetic spectrum, 29
electromagnetic waves
    light waves, 15
    movement of, 16–20
    properties of, 6, 14–16, 18, 25,
        27, 28–29, 38
    radio waves, 14–15, 27–28
energy transport phenomenon
    (elasticity), 11–13, 17
equilibrium position
    (rest position), 8,
        22, 37

F
frequency, 26–27, 29, 37–38

H
Hertz, Heinrich Rudolf, 27–28
hertzian waves, 27–29
Huygens, Christian, 16

I
interference, 36–37
    constructive, 36–37
    destructive, 36, 37

L
longitudinal waves, 17, 18–20,
    22, 25, 41
    compressions, 19, 22
    rarefactions, 19, 22

M
magnitude, 24–25
Marconi, Guglielmo, 28
Maxwell, James Clerk, 27
mechanical waves
    medium, 9–11, 15, 16, 17, 18,
        19, 30
    movement of, 16–20
    properties of, 6, 7–9, 14, 15, 16,
        18, 25, 38

N
National Radio Astronomy
    Observatory (NRAO), 41–42
Newton, Sir Isaac, 16

O
ocean wave, 14, 19–20

P
pascal, 22
period, 25
Planck, Max, 16

R
radio astronomy, 41–42
reflection, 30, 32–34
    law of, 34

refraction, 30, 32, 34–35
ripple tank, 31–32, 33,
    34–35
Robert C. Byrd
    Green Bank Telescope
    (GBT), 42

S
seismology, 40–41
sonar, 5, 39–40
sound waves, 15, 19,
    37–38, 40
spectrum analyzer, 29
surface waves, 17, 20

T
transverse waves, 17–18, 19, 20,
    21–22, 25, 41
    crests, 21–22, 26, 36
    troughs, 21–22, 26, 36
tsunami, 11, 34

V
vacuum, 7, 14, 15–16, 27
velocity, 37

W
wavelength, 22, 25, 29, 37, 38

About the Author

April Isaacs earned an MFA degree in nonfiction creative writing from New School University. She was born in Fort Wayne, Indiana, and currently lives in New York City.

Photo Credits

Cover © Mehau Kulyk/Photo Researchers, Inc.; pp. 5, 10 (top left, top right, bottom) © AP/Wide World Photos; pp. 8, 18, 20, 26, 31, 33, 35, 36, 38 by Geri Fletcher; p. 112 © 2000–2004 Custom Medical Stock Photo; p. 15 © Oscar Burriel/Photo Researchers, Inc.; p. 16 © S. Terry/Photo Researchers, Inc.; p. 23 © Bettmann/Corbis; p. 28 © Hulton-Deutsch Collection/Corbis; p. 40 © Reuters/Corbis; p. 42 © Dr. Seth Shostak/Photo Researchers, Inc.

Designer: Tahara Anderson; **Editor:** Kathy Kuhtz Campbell

www.ingramcontent.com/pod-product-compliance
Lightning Source LLC
Chambersburg PA
CBHW050911210326
41597CB00002B/94